THE LIGHT

LIVING WITH MODERN CLASSICS

THE LIGHT

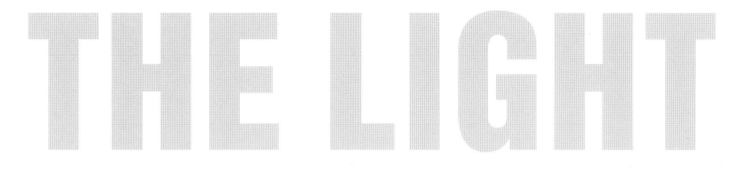

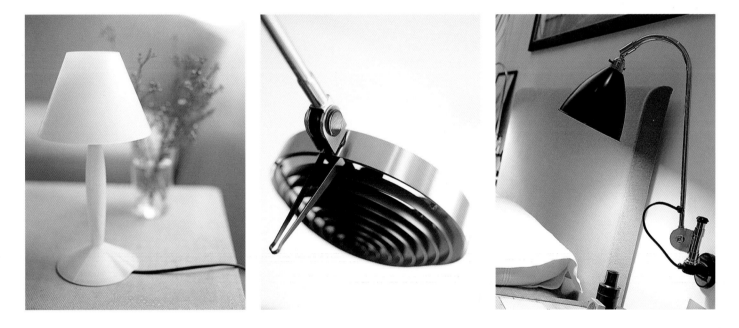

Elizabeth Wilhide *with photography by* **Chris Everard**

WATSON-GUPTILL PUBLICATIONS / NEW YORK

First published in the United States in 2000
by Watson-Guptill Publications, a division of
BPI Communications, Inc., 1515 Broadway,
New York, NY 10036

Library of Congress Card Number: 00–100629

ISBN: 0-8230-3110-1

This book was conceived, designed, and produced
by Ryland Peters & Small
Cavendish House
51–55 Mortimer Street
London W1N 7TD

Senior designer Ashley Western
Senior editor Annabel Morgan
Location researcher Kate Brunt
Stylist Christina Wilson
Production Patricia Harrington
Art director Gabriella Le Grazie
Publishing director Alison Starling

Printed and bound in China

First printing, 2000
1 2 3 4 5 6 7 8 / 07 06 05 04 03 02 01 00

CONTENTS

Right *A collection of modern lights in an open-plan loft space: from left to right, the Fortuny light, Vico Magistretti's Chimera, and the Panthella by Verner Panton.*
Below *Bulb: Ingo Maurer's witty homage to Thomas Edison, the inventor of the incandescent bulb.*

Introduction

The electric lightbulb is the classic product of an ingenious age. Although the same idea occurred to a number of nineteenth-century inventors and several different designs were patented, it was Thomas Edison who first addressed the practicalities of the subject and made the revolutionary new development a commercial success. Offering for the first time ever a source of illumination that was clean, even, and easy to use, Edison's incandescent bulb, patented in 1879, was a key factor in stimulating a demand for electricity in the home.

Despite the advantages of electric lighting, it was not until the twentieth century, when a power supply infrastructure had been established and the cost of electrifying homes had come down, that its use became widespread in nearly all sectors of society. During this time of transition—the period immediately following the invention up until the dawn of the machine age—the electric light took two main forms.

The first—and shortest-lived—was the simple exposure of the bare incandescent bulb. In those early days, when electricity was both a novelty and a luxury, electric bulbs were most often hung unshaded, even in the most richly and ornately decorated surroundings. This arrangement, which would strike many people as oddly utilitarian today, was no doubt intended as a proud display of the latest technology. Bare bulbs hung around the perimeter of the lavishly appointed New York drawing room of millionaire financier J. Pierpoint Morgan (1883), while in Britain, Cragside, the Northumberland mansion home of the wealthy arms manufacturer Lord Armstrong, was the first house in that country to be electrified. Cragside had its own generator.

While early wattages were dim by today's standards, to a generation accustomed to soft, flattering, and atmospheric candlelight, the electric bulb could seem very glaring indeed. A common complaint was that electric

light destroyed any sense of intimacy: it robbed rooms of "privacy and distinction" in the words of the American novelist Edith Wharton, whose first published works were on the subject of interior decoration. Accordingly, a mere 20 years after the Edison bulb was patented, there was already a growing light-fixtures industry. Such fixtures were designed to shade the light source—to dim it—and to provide the consumers of the new technology with the reassurance of tradition by incorporating elements that suggested the candlestick, gas mantle, or lantern. By 1905, Tiffany employed 200 craftsmen to produce stained glass pendants and lamps in the newly fashionable Art Nouveau style. In its second manifestation, the electric light was transformed into a decorative object, a familiar and predictable disguise for a functional element.

In the early decades of the twentieth century, a new wave of radical designers and architects proposed an abrupt break with such traditions. Modernists, with their familiar mantra "form follows function," took inspiration from the machine, and from industrial processes and materials. One of the earliest lights included in this book, Wilhelm Wagenfeld's WG 24, was a product of the Bauhaus, perhaps the seminal school of design of the twentieth century. Incorporating industrial-style elements and expressing the modern aesthetic in its use of glass and steel tubing, the light was uncompromisingly utilitarian for its time. Eileen Gray's minimal Tube Light was even simpler and revealed the modernists' love affair with tubular forms. The bulb was no longer an ugly necessity that needed to be disguised, but part of a working machine whose purpose

was to deliver light. In this context, nothing could be more explicitly mechanical than the Anglepoise, the jointed desk lamp designed by George Cawardine in the 1930s.

As demonstrated by the selection of work presented in this book, since World War II Italian designers have dominated the field of lighting design. In the postwar reconstruction of the Italian economy, a particularly fruitful collaboration arose between designers and manufacturers. Light fixtures, which did not require huge capital investment to produce but which could, through the value added by design, attract a luxury market, made perfect export products. Architects and designers worked as consultants for established firms or, in many cases, set up their own companies to exploit new technologies and new materials. In the process, the relationship between lighting design and technological advance became even closer.

The work of Achille Castiglioni and his brothers Livio and Pier Giacomo reveals just such an interest in the technological cutting edge, married to a playful approach that derives from the explorations of Marcel Duchamp and his "ready-mades." The Toio, comprising a car headlight, a fishing rod, and a bandsaw, is a mechanical patchwork, one of the first designs to make use of a halogen light source. Years before "hi-tech" became fashionable, Castiglioni's designs predicted the influence that specialized forms of lighting—in cars, retail displays, and photography—would come to exert on design for the home. In the work of other contemporary designers, light fixtures grew ever more elegant, more minimal, the ultimate in covetable "designer objects." Lights such as the Tizio desk lamp and the Jill uplighter reveal the Italian desire to create lights that have a sculptural presence even when not in use.

With the advent of low-voltage halogen in the 1980s, the trend for miniaturization accelerated. Tiny halogen bulbs poised on tensioned bare wire reduced the light to

Opposite *The Brera light, with its pure ovoid shape, is a recent design by the prolific and hugely influential Italian lighting designer Achille Castiglioni.*
Left *The jointed Anglepoise is a lighting classic, originally designed in the 1930s by British engineer George Cawardine.*
Below *Wagenfeld's simple table lamp, WG 24, was a product of the Bauhaus school of design.*

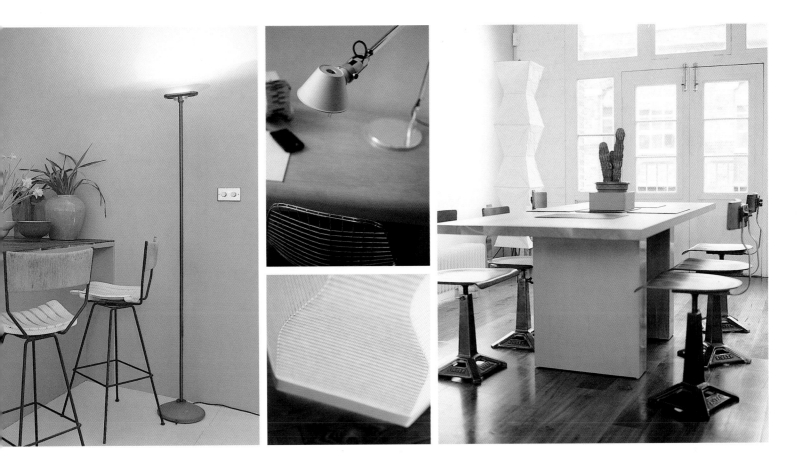

Opposite Tom Dixon's Jack doubles as an occasional seat.

Above left The Jill uplighter is an iconic design of the late 1970s.

Above center Tolomeo, by Michele de Lucchi and Giancarlo Fassina, is an acclaimed Italian classic.

Above center below Simmons' recycled cardboard Lumalight.

Above right Noguchi's Akari lantern is one of the most appealing of all contemporary designs.

the barest of essentials. The clarity and true color rendering of halogen, which had made it invaluable as a light source in the retail environment, was increasingly in evidence in the home, replacing the muted golden tones of the old tungsten bulb.

At the same time, a new direction could be detected in lighting design—one that was more expressive, more emotive, and sometimes even positively low-tech. Tom Dixon's Jack, for example, is an object or seat that just happens to light up; Ingo Mauer's poetic Lucellino is a light with tiny feathered wings; Philippe Starck's dreamy Rosy Angelis shrouds the standard lamp in mystery. In

such designs, the level of technological sophistication is most definitely not the issue. The heart of the matter is light itself and its unique animating presence.

One designer who appreciated the captivating qualities of light better than most was Isamu Noguchi. Nothing could be simpler or more evocative than his Akari lanterns; few designs have been more imitated. Noguchi's great concern was the quality of light, and the soft intimacy of his light sculptures in paper and bamboo have proved perennially appealing. Perhaps it is no surprise that, after a century of experimentation, the plain paper lantern remains one of the most popular light fixtures of all.

ROBERT DUDLEY
BEST 1892–1984

The designer of the Bestlite collection, Robert Dudley Best, studied at the Dusseldorf School of Industrial Design and the Interior Design Atelier in Paris. Absorbing the radical teachings of the Bauhaus and influenced by the 1925 International Exhibition of Modern and Industrial Arts in Paris, where the work of Mies van der Rohe and Le Corbusier was prominently displayed, he went on to create one of the most enduring classics of modern lighting design.

Best & Lloyd, the British company for which Best's designs were produced, was first established in 1840. In addition to manufacturing the Bestlite, the company specializes in individually crafted brass light fixtures.

Bestlite (1930)
Manufacturer: Best & Lloyd

Available in a lighting collection that includes a desk lamp, table lamp, wall-mounted lamp, floor lamp, and cantilever lamp, the Bestlite displays all the clarity characteristic of early modernism. The shade, on its gently curved arm, rotates and tilts through its axis. Similarly, the arm tilts up and down, while in the table and floor lamp versions, the entire assembly can be fully rotated around the supporting pillar or slid up and down to assume any position. This universality of movement makes for a highly functional light, ideal for any situation in which bright, direct light is required.

In continuous production since it first appeared in 1930, the Bestlite's enduring status as a modern classic owes a great deal to the extremely superior quality of its manufacture. Each light is handmade, and different materials and finishes can be specified.

Above The superbly crafted Bestlite is available in a choice of formats, including the table-lamp version.
Right The floor-standing Bestlite is fully adjustable. The arm that holds the shade, slides up and down the stem and rotates. The shade can be angled to change the direction of the light. Opposite The wall-mounted Bestlite makes a perfect directional task light for illuminating work areas, countertops, or bedsides.

SANTIAGO
CALATRAVA b. 1951

Born in Spain, but currently based in Switzerland, Calatrava is a prolific and
expressive designer whose work ranges from engineering masterpieces such as
Puente del Alamillo, the bridge he designed for the Seville World's Fair in 1992,
to a train station in Lyon, France (1994). Trained as both architect and engineer,
Calatrava is perhaps best known for his bridges, including those at Valencia and
Barcelona, which typically combine technological sophistication with an organic
sensibility. At Puente del Alamillo, a strikingly asymmetric design, a huge
concrete pillar clad in steel supports 13 pairs of cables that hold up the deck
of the bridge. The sculptural quality of the pillar, like the steel arches that
comprise the station roof at Lyon, betrays the designer's interest in natural forms.
He is reputed to keep the skeleton of a dog in his office for design inspiration.

Montjuic (1990)
Manufacturer: Artemide

This iconic floor light echoes the dynamism of Calatrava's great
engineering projects. The swept-back angle of the synthetic resin
stand suggests tension, as if the lamp itself were a missile poised for
release. The splayed feet counterpoint the shape of the sandblasted
glass shade. From the bonelike profile of the stand to the prongs that
grip the shade like claws, the organic, natural references are plain
and serve to animate what is essentially a simple design.

   While the light functions basically as an uplighter, the diffusing
nature of the shade softens the light source, creating an effect
reminiscent of a traditional torch-holder.

This page and left The Arco is a seminal design of the 1960s. Although it was originally intended as a means of lighting a dining table, it is equally popular as a dramatic focal point in a living area. The heavy base that anchors the design is made of marble.

ACHILLE and PIER GIACOMO
CASTIGLIONI
b.1918 and 1913–1968

The two youngest of three Milanese brothers, both Achille and Pier Giacomo Castiglioni studied architecture at Milan Polytechnic. After graduating in 1944, Achille joined the design studio set up by Pier Giacomo and their brother Livio (1911–79). The creative collaboration between Achille and Pier Giacomo, in particular, made a significant contribution to design in postwar Italy.

Sons of a sculptor, the conceptual approach of the Castiglioni brothers owes much to Marcel Duchamp's "found objects" or "ready-made" pieces. One of their most famous early designs was the Mezzadro stool (1957), a tractor seat on a bent steel stand. But their towering reputation for ingenuity and wit derives from their work in the field of lighting design. From their early experiments with industrial lamps to the Arco lamp of 1962, they established a distinctive identity, a unique blend of humor and functional elegance. Following Pier Giacomo's death, Achille has continued to produce a wide range of designs for leading Italian manufacturers: flatware, glassware, chairs, and tables, as well as lights.

Arco (1962)
Manufacturer: Flos

A design that defines the chic modernism of the 1960s, the Arco floor lamp was one of the most successful results of the design partnership of Achille and Pier Giacomo Castiglioni. A persistent theme of their work was the juxtaposition of different materials, here expressed in the combination of the arching stainless steel stem and monumental white marble base. Again, the contrast between the springiness of the arc and the solidity of the base generates a sense of animation.

The Arco was originally designed as a way of lighting a dining table without relying on a permanent overhead fixture and can therefore be adjusted to three different heights, with the arc broad enough to allow movement around the table. The reflector shade is made of polished aluminum. The steel stem serves as a channel for the cord.

Toio (1965)

Manufacturer: Flos

First produced in 1965, Toio provides a good example of the Castiglioni brothers' creative reuse of ready-made objects. A good decade or more before "hi-tech" became fashionable, the Toio uplighter represented a radical departure in the field of home lighting. The design came about when the Castiglioni brothers were challenged to create a light from random items lying about in a garage. The original combined a fishing rod for the stand, a band saw as the brace, and a halogen car headlight as the light source.

    The enameled steel and brass stand of the lamp is extendable. All Toio's working parts, including the lamp, cabling, and heavy transformer, are left defiantly exposed.

Parentesi (1970)

Manufacturer: Flos

One of Achille Castiglioni's best-known lights, Parentesi is the result
of a collaboration with Pio Manzù (1939–69), a designer employed by
the Fiat Design Center and responsible for the concept drawings of the
Fiat 127. The Parentesi, which makes use of car reflector bulbs,
reveals both Manzù's background in the automobile industry and
Castiglioni's lifelong fascination with the ready-made. Unfortunately,
Manzù was killed in a car accident before the design was completed.

Simple and functional, the lamp is attached to a metal sleeve that
can be moved along a steel cable suspended between ceiling and
floor. The position and direction of the lamps are adjustable in any
direction, and light levels can also be varied in intensity.

Brera (1992)

Manufacturer: Flos

The Brera light has a timeless quality that derives from its inspiration, an ostrich egg. The egg in question appears in Piero della Francesca's Montefeltro altarpiece, in which it is suspended in midair over the head of Mary—an ancient symbol of purity denoting the virgin birth. The light takes its name from the Pinacoteca di Brera in Milan, where the altarpiece is now on display.

Made of frosted glass and available in several formats—as a pendant, table, wall, ceiling, or floor-standing lamp—the Brera demonstrates Castiglioni's enduring powers of creativity: he was 74 when he designed it.

Frisbi (1978)

Manufacturer: Flos

A spare, elegant pendant fixture, the Frisbi has a "look-no-hands" quality that adds dramatic impact. Avoiding glare is the principal problem to address in the design of pendant or overhead lighting. Castiglioni's classic solution comes in the form of a glass disk, suspended from barely visible fine wires so that it hangs a short distance beneath the metal light fixture. The glass disk provides both a soft diffusion of light and a more concentrated, directional beam through the hole in its center. As the name suggests, this highly sculptural design looks poised for movement.

Far left *Inspired by the frisbee, as its name suggests, this Castiglioni pendant provides unobtrusive glare-free lighting for a dining table.*

This page *Castiglioni was in his mid-70s when he designed the Brera, a light whose form was inspired by an ostrich egg depicted in a painting by Piero della Francesca.*

GEORGE CAWARDINE 1887-1948

George Cawardine was an automotive engineer, and director of Cawardine Associates of Bath, when he designed one of the most enduring design classics of the twentieth century. The Anglepoise, first produced more than 60 years ago, has proved one of the most popular functional lights in Britain, with "anglepoise" becoming virtually a generic term for this type of adjustable desk lamp. The light was designed for the British firm Herbert Terry and Sons.

Cawardine had an abiding interest in ergonomics and famously based his classic design on the anatomy of the human arm.

Anglepoise (1932)

Manufacturer: Anglepoise

A familiar sight in design studios, offices, and homes, the Anglepoise is supremely functional, delivering bright directional light to the desk, bedside, or drawing board. Springs act as a counterweight to hold the arm of the lamp in position, and the mobile joints mean the light can be angled at a variety of heights.

In the original design, the shade and arm of the lamp were made of lacquered metal, while the base was made of Bakelite, an early form of plastic. Today, the Anglepoise is generally all-metal, available in a wide variety of colors and with either a solid circular base or a screw clamp to allow it to be mounted on the side of a desk or table. Outside Britain, the best-known version of the design is the Luxo light, produced by the Danish manufacturer Jacob Jacobsen after he acquired the Anglepoise patent in 1937.

Right One of the most popular functional lights ever designed, the Anglepoise is also one of the most imitated.
Left The Luxo, a simplified Danish version of the original, is available in a variety of colors.

TOMASSO CIMINI b. 1947

Lighting designer Tomasso Cimini began his career as an engineer working for
the Italian lighting manufacturer Artemide before founding his own company,
Lumina, in 1978. One of his best-known designs is the Daphine lamp, produced
for Lumina; another Cimini design for Lumina, Zed (1987), is a futuristic table
lamp, with small halogen accent spot at the end of a sinuous snakelike arm.

Elle (1986)
Manufacturer: Lumina

**This floor-standing accent or directional light is an essay in
attenuation. The jointed stand has an almost precarious look, as if it
might topple over at any moment. With the round bright "eye" of the
halogen lamp supported by the spindly metal stand, there is a trace
of anthropomorphism in the design, giving it character and humor.**

DEVESA I BAJET

Representative of a new generation of Spanish design talent, the Devesa brothers were born in Barcelona—since the mid-1980s the undisputed design capital of Spain—and trained as designers in the city, studying at the School of Arts and Crafts, School of Design, and School Elisava. After leaving college in 1986, they started to work with the Spanish firm Metalarte on new lighting projects and also undertook the design of Metalarte showrooms in Barcelona, Valencia, and Paris.

The following year saw the Devesas launch their own company, D&D Design, producing designs for a number of national and international companies, including Disform, Amat, and Blauet, as well as Metalarte, for whom they have created a number of sleek minimalist wall lights. Their work has featured widely in design competitions and exhibitions, including the SIDI Design Fair, the Nuevo Estilo design awards, and the Design in Catalunya exhibition, 1987.

Zen (1990)

Manufacturer: Metalarte

The Devesas have produced a number of lights for Metalarte, of which the Zen is probably the best known. Made from polished cast aluminum, this compact table lamp is tiny—a mere 6¼ inches tall and 4 inches across the base—although it is surprisingly heavy. Like much of the Devesas' work, there is a decidedly retro element to the design, with the metal grille and rounded form recalling the streamlined shape of early twentieth-century appliances, while the small scale adds to the light's appeal. The use of sleek metallic finishes and matt glass is typical of the Devesas' work.

This page and opposite

Utterly covetable, the tiny Zen table lamp is light as designer object. The retro elements of the design, including the integral switch, give it the appearance of an old-fashioned appliance—or an early microphone. When the lamp is lit, the metal grid glows like a honeycomb.

TOM DIXON b. 1959

One of the leading figures in the new wave of British design talent, Tom Dixon was born in Tunisia, but returned to live in England at the age of four, growing up in Huddersfield and London. With his interest in both art and design, it was a natural progression for him to enroll in a degree course at the acclaimed St. Martin's School of Art in London. However, Dixon soon found the organization of the course too rigid and restrictive, particularly the fact that students were required to specialize, and he left after six months.

In the early 1980s, after he left college, Dixon set up Creative Salvage with other ex-students, producing furniture comprising "ready-made" elements such as railings, pans, and tubing welded together. His next venture was Space, a store selling furniture and lighting designed by himself and other contemporaries; Dixon's famous S chair dates from 1988. In 1996, Dixon sold Space and set up Eurolounge, which produces designs in plastic. An opportunity to bring his ideas to a wider market came in 1998 when he was appointed Design Director of Habitat. Like Sir Terence Conran, Habitat's original founder, Dixon has a declared interest in high-quality affordable design and has introduced modern classics by designers such as Verner Panton, Robin Day, and Achille Castiglioni to Habitat's line.

Jack (1996)

Manufacturer: Eurolounge

Refusal to be bound by conventional categories characterizes Tom Dixon's work. Like the S chair, a sinuous wicker-clad curve, Jack is both sculpture and design; furthermore, it is both a light and a seat. Made of polypropylene, the design is in the form of an over-scaled jack and can be displayed singly, grouped, or even stacked, and is robust enough to serve as an occasional low level seat. Available in a choice of colors, this optimistic, witty design—which has been a great commercial success—reflects the growing trend among designers to treat the light increasingly as a sculptural, rather than simply a strictly functional, object.

When is a light not a light? When
it is also a seat—or even an
overscaled jack made of smooth
plastic. The soft, rounded curves and
ambient glow of this Tom Dixon design
add wit and humor to the interior.

MARIANO FORTUNY Y MADRAZO
1871–1949

The name Fortuny is most strongly associated with the fine pleated silk dresses that were so popular among fashionable artistic women in the 1920s. But in fact, Mariano Fortuny, who was born in Grenada, Spain, raised in Rome and Paris, and who moved to Venice in 1889, began his career in stage design and photography. The Fortuny light dates from this early period.

In the second half of his career, Fortuny became interested in experimenting with printed and pleated silk, and in fashion design. From 1907 on, he designed the classically inspired dresses for which he became so famous and opened a factory in 1919 to produce the pleated material. Fortuny's luxurious gowns, which clung sensuously to the body and were virtually crushproof, were supposedly so fine they could be slipped through a wedding ring.

Fortuny (1907)

Manufacturer: Pallucco Italia

Fortuny's several interests—in the theater, photography, and fabric— are jointly expressed in the design of the Fortuny light, which was originally manufactured under an arrangement with the German company AEG. Astonishingly modern for its time, this theatrical light stands on a sturdy steel tripod; light from the lamp bounces off an overscaled fabric hemisphere and a reflective metal panel, creating a bright yet diffused light, free from dazzle and glare. The arrival of "hi-tech" in the 1970s witnessed the increasing popularity of technical, theatrical, and photographic lights in domesticated settings.

The name Fortuny may be more
readily associated with couture
dresses, but the versatile designer
was also responsible for this classic
'photographer's' light. The cotton
hemisphere acts as a diffuser.

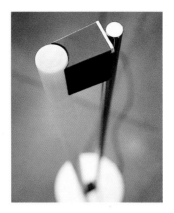

Left, right, and opposite

Eileen Gray is perhaps better known for her rugs and textiles—and for the Bibendum armchair—than for lighting designs. But the Tube Light displays her characteristically elegant modernism. The functional tubular form, itself a preoccupation among early modernists, is here transformed into a glowing wand.

EILEEN
GRAY 1878–1976

Eileen Gray produced a wide variety of designs for furniture and textiles that managed to marry functional modernism with a certain sense of chic and sophistication. Born in Ireland into an artistic, aristocratic family, she studied first at the Slade School of Art in London in 1898 and then moved to Paris where she was tutored in the art of lacquerware by the Japanese master Sugawara. Her early work in Paris was highly decorative and often displayed a number of Oriental motifs. She began designing interiors in the early 1920s, after opening the Galerie Jean Désert to sell her own furniture.

The influence of modernism can be detected in Gray's work of the mid-1920s. Fascinated by the potential offered by new materials, such as tubular steel, and increasingly drawn to strong geometric forms, her characteristic style was first displayed in the villa she designed in the south of France with Jean Badovici. The villa, E-1027, was furnished entirely with her own textiles, carpets, and furniture designs, which included the luxurious Bibendum armchair, one of her most famous and celebrated designs. Although her architectural work was exhibited in Le Corbusier's pavilion for the 1937 Paris Exhibition, until relatively recently she lacked the prominence of other modernist designers.

Tube Light (1927)

Manufacturer: ClassiCon GmbH

Dating from the period when Gray was working on the design of her own villa in the south of France, her Tube Light displays the pared-down simplicity that is so characteristic of early modernism. At the same time, the sheer elegance of this slender wand of light betrays Gray's sympathetic and sensitive interpretation of a style that could in other hands be almost brutal in its hard-edged rationality. The design, which is essentially a tube supported by a slim chromium-plated steel frame, serves as a welcome vertical counterpoint to the long, low lines so typical of modernist interiors.

POUL HENNINGSEN 1894–1967

Poul Henningsen studied architecture at the Technical University of Denmark in Copenhagen, and set up in practice in 1920. Although his architectural work included houses, restaurants, and theaters, he is without a doubt now best known for his lighting designs, which first won critical acclaim when they swept the board at a Danish lighting competition in 1924, taking all the prizes. In 1925, Henningsen's lights were exhibited in the Danish Pavilion at the Exposition Internationale des Arts Decoratifs in Paris, where they won a gold medal.

With the PH series, designed in 1926 for the Forum, a Copenhagen exhibition hall, Henningsen's lights became a popular as well as a critical success. Mass-produced by Louis Poulsen, the Danish lighting manufacturer, they became a familiar feature of many progressive Scandinavian interiors and are still in production today. Later versions of the designs were based on the same basic principle: using overlapping planes to diffuse the light source and avoid glare.

In the late 1920s, Henningsen co-edited the influential Danish architectural magazine *Kritisk Revy* with the architect and designer Kaare Klint (*see pages 40–41*). A key figure in the establishment of the "Scandinavian Modern" aesthetic, he was a passionate advocate of the integration of art into manufacturing.

This page and opposite

With their natural sensibility and organic form, the PH series of lamps encapsulate the humane modernism of Scandinavian design. There are a number of different versions, including pendants, hinged pendants, and a table lamp.

PH series (1926 on)

Manufacturer: Louis Poulsen

The modern PH pendant (credited to Poul Henningsen, Ebbe Christensen, and Sophus Frandsen) is a slight reworking of the PH 5, designed in 1958, itself a reworking of the original PH hanging lamp of 1926. Made of white-painted aluminum, rather than the opaque glass of the original, the fixture displays Henningsen's concern with the quality of light. Like many Scandinavian designs, the modernity of the design is tempered by an organic quality, with the metal planes or "leaves" that diffuse the light suggesting a plant form. Other lamps in the PH series include wall and table versions, while the PH-Zapfen or "Artichoke" is an elaboration of the basic principle.

Left and opposite A design that is virtually unimprovable, like much of Jacobsen's work, the discreet AJ visor is simplicity itself. It is a simplicity, however, that is never simplistic, but is the result of supreme refinement. The angle of the shade and the minimalism of the stand and stem give the design a look of alertness and focus. **Right** The design is available in a number of different finishes and in a table lamp version.

ARNE JACOBSEN 1902–1971

Born in Copenhagen and trained as an architect at the Royal Danish Academy of Fine Arts, Arne Jacobsen went on to become one of the most famous exponents of Scandinavian modernism. At the Paris Exhibition of 1925 (where one of his chairs won a medal), Jacobsen was exposed to the work of Mies van der Rohe and Le Corbusier; after establishing his own practice in 1930, he introduced the modernist aesthetic to Denmark, albeit characteristically tempered by a very Scandinavian sense of craftsmanship and sympathy for materials.

Throughout a long and successful career, Jacobsen designed everything from buildings to flatware, furniture, textiles, and light fixtures—a breadth of approach that reflected his belief that buildings and what they contained should be conceived as an integrated whole. In this respect, the Royal SAS Hotel, Copenhagen (1958), and St. Catherine's College, Oxford (1960–64), can perhaps be seen as his most significant achievements. The success of designs such as the 3107 chair (1955) and the Swan and Egg chairs (1957), originally created for the Royal SAS Hotel, brought his work to a wider public. Jacobsen-designed flatware was selected by Stanley Kubrick for his film *2001: A Space Odyssey*.

AJ Visor (1956)

Manufacturer: Louis Poulsen

This beautifully simple light demonstrates that the functional can also be supremely elegant. The lamp can be adjusted to provide concentrated directional lighting for reading or study; the metal shade is steeply angled, like a visor, to shield the eyes from the glare of the light source. The base, with its circular cutout, serves as a counterbalance to the stand. The overall effect is one of poise and refinement. The AJ Visor is available as both a table and floor lamp.

British-born designer Perry King studied at Birmingham College of Art before moving to Italy in 1965 to take up a post as consultant to Olivetti. There, he collaborated with Ettore Sottsass on a number of projects, including the Valentine typewriter (1969), with its bold pop styling, and the Synthesis 45 office furniture system. King became design coordinator of Olivetti's corporate identity department in 1972.

In 1977, King established a studio in Milan with Spanish-born Santiago Miranda. King-Miranda has since designed graphics, corporate identity programs, furniture, and office equipment. Lighting, however, is a particular interest. In 1996, King-Miranda was the first non-Scandinavian design house to be commissioned by Danish manufacturer Louis Poulsen, for whom they created the Borealis outdoor light, now widely used in public areas. Giancarlo Arnaldi has collaborated on many of King and Miranda's lighting designs.

This page and opposite

The hugely popular Jill uplighter has been designed to appeal just as much when the light is switched off as when it is in use.

Jill (1978)

Manufacturer: Flos, Arteluce

An extremely popular design, Jill makes a sophisticated statement in both classic and contemporary interiors. The circular base and reflector shade are made of colored opaque glass; the stand is metal. The shallow glass shade both diffuses the light and directs it upward. Jill was one of the first lighting designs to exploit the potential of tungsten-halogen, a bright, clean white light source that lends itself to use in uplighters.

The stepped profile of the reflector has the faintest hint of Art Deco: this is a lamp that has been designed both to function effectively and to appeal as a decorative object when not in use. The combination of high technical performance and attractive, evocative packaging is distinctive of King-Miranda's work.

KAARE
KLINT 1888–1954

P.V. Jensen Klint, a teacher at the Technical School in Copenhagen, first began producing his own paper shades for personal use around the turn of the century. Friends and family joined him in experimenting with a variety of production techniques and contributed designs of their own. By the 1940s, demand far outstripped supply, and a family firm, named Le Klint, was set up to produce the lampshades commercially. The first design intended specifically for mass production, the Fruit shade, was designed by Jensen Klint's son, the influential architect and designer Kaare Klint, in 1944.

Kaare Klint went on to design a number of paper shades for Le Klint, many of them in collaboration with his son, the industrial designer Ebsen Klint. Like much of Klint's work, the shades betray a fundamental taste for simplicity; unlike his furniture they were specially designed for mass production. Klint was inspired by the skills of the great eighteenth-century English cabinetmakers, and most of his furniture, in direct contrast to the Le Klint shades, was crafted by hand.

This page and opposite

Cheap, readily available, and an
excellent diffuser of light, paper
makes an economic and effective
light shade. These plastic-coated
pleated shades manufactured by Le
Klint are a modern Danish classic.

Paper shades (mid-1940s on)

Manufacturer: Le Klint

Simple, elegant, and inexpensive, the folded and pleated paper
shades manufactured by the Le Klint family firm have remained in
continuous production since World War II. Nowadays, the shades are
plastic or plastic-coated to promote durability and ease of cleaning.
The intricate folds and pleats are scored onto the plastic by machine,
but most shades are still hand assembled.

In the late 1940s Le Klint discovered that the designer Poul
Henningsen *(see pages 34–35)* had developed some very similar
designs for the Danish lighting manufacturer Louis Poulsen. A court
case ensued, which Le Klint won; ever since, the company has had
exclusive rights to market the design.

This page and opposite

*A beautiful working object, the
Tolomeo's obvious antecedent is
the jointed desk lamp, here updated
in lightweight brushed and polished
aluminum. A neat feature is the hole
pierced at the rear of the shade; the
ribbed base provides another
example of attention to detail.*

MICHELE DE LUCCHI b. 1951
and GIANCARLO FASSINA b. 1935

Michele de Lucchi began his career as an industrial designer, working for well-known Italian design companies such as Artemide, Olivetti, and Kartell. But it was his association with the Milan-based Memphis group, led by Ettore Sottsass, that won him wider recognition in the 1980s.

De Lucchi, like his fellow Italians, the Castiglioni brothers, has had a long interest in the exploitation of "ready-mades" in design, a direction that ultimately derives from the early twentieth-century work of Marcel Duchamp, who famously shocked and provoked the art world when he put a urinal on display and called it art. More recently, de Lucchi has been involved with the creation of an experimental line of products, available in limited editions, which make use of "found objects." His Treforchette Table Lamp (1997), for example, consists of a cylindrical plastic lamp shade supported by three ordinary table forks.

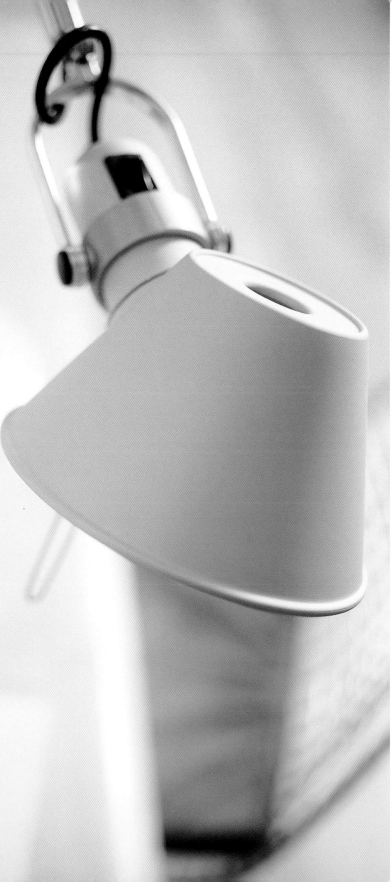

Tolomeo (1987)

Manufacturer: Artemide

A particularly pleasing variation on the theme of the jointed desk lamp, the Tolomeo is a modern classic of Italian design. Made of lightweight aluminum, either oxidized or painted black, the design is also available in both floor and wall-mounted versions.

Tolomeo's small flowerpot shade delivers a bright pool of concentrated light to the work surface. An unobtrusive metal handle attached to the rim of the shade allows the light to be repositioned comfortably without the need to touch the heat of the shade.

VICO
MAGISTRETTI b. 1920

Architect and product designer, and one of the most influential figures in the creation of the postwar Italian design identity, Vico Magistretti was born in Milan, trained as an architect in his native city, and went to work in his father's architectural studio after graduating in 1945. During the postwar period of reconstruction, Magistretti, like many other Italian designers, came to view collaboration between design and manufacturing as offering the greatest potential for future economic regeneration. Early designs for low-cost furniture were followed in the 1960s by a long association with progressive Italian companies such as Cassina and Artemide. His Carimate chair for Cassina (1964) became an iconic design of the 1960s. With its glossy, brightly colored wooden framework, rush seating, and strong simple lines, it represented a youthful take on a traditional Italian rustic style and was much imitated. Another popular design was the Atollo lamp, produced by O Luce. This lamp, with its elegant geometric lines, was winner of the 1979 Compasso d'Oro prize for product design.

In addition to products such as chairs, tables, and lights, Magistretti has also continued to design buildings. He has received many awards and has been an influential teacher of design in Milan, Tokyo, and London.

This page and opposite

Lighting design has been informed by the development of new materials as much as by the development of new technologies. The sensuous curves and glossy finish of the Chimera give a new twist to plastic, once considered a cheap and disposable material.

Chimera (1966)

Manufacturer: Artemide

Like the Selene chair (1969), which Magistretti also designed for
Artemide, the Chimera helped to redefine the image of plastic,
previously seen as a cheap, disposable material. Here, the glowing,
rippling planes of this floor-standing lamp give the material an
almost luscious quality. The soft, ambient light diffused through the
scroll of plastic transforms the lamp into a sculptural object.

INGO MAURER b. 1932

One of the most consistently inventive of all lighting designers, German-born Ingo Maurer exploits new technology to create products that are both expressive and playful. Mauer studied typography in Germany and Switzerland (1954–58), then spent three years working as a freelance designer in the United States, where he was much influenced by the "Pop Art" movement. In 1966 he established his studio, Design M, in Munich.

Mauer's early lighting designs betrayed his interest in pop. Bulb (1966), for example, consisted of an ordinary lightbulb within a large clear bulb. But it was his experiments with cutting-edge technology that first won him international acclaim. The Ya Ya Ho lighting system (1984) was one of the first bare-wire lighting installations, exploiting the development of low-voltage halogen. Small halogen lights, attached to tensioned cabling and counterbalanced by weights, create the effect of a mobile. The light sources can be rearranged for a variety of effects. In all of Maurer's work, however, such technical sophistication is merely a means to an end: the result inevitably blurs the boundary between art and design.

Lucellino (1992)
Manufacturer: Ingo Maurer Gmbh

One of Maurer's most poetic flights of fancy, Lucellino is a light with wings. Maurer is one of several designers who view the lightbulb as a beautifully designed object in its own right. Here, a bare 24-volt bulb is given goose-feather wings and supported by a copper wire around which a red electric cord is entwined. The wire is pliable, so the light can be angled in any direction; touching the wire creates a resistance that dims the light. The light comes with a transformer.

Bulb (1966)

Manufacturer: Ingo Maurer Gmbh

One of Maurer's earliest designs, Bulb makes reference to pop culture. Intended as a personal homage to Thomas Edison's "bright idea," the design celebrates the incandescent bulb, the nineteenth-century invention that completely transformed household lighting, and which is now so ubiquitous as to be almost unregarded. The light source is a crown-silvered 100-watt bulb, which is encased in a glass lightbulb of its own. The base is chrome-plated.

ALBERTO MEDA b. 1945

Italian-born designer Alberto Meda is an engineer by training, and it is this background that has given him an intense interest in the technology of materials. Meda studied engineering in Milan, graduating in 1969, and worked as an assistant at Magneti Marelli until 1973. For the next six years, he was technical director at Kartell, a Milan-based company engaged in the research and development of plastics. Meda set up as an independent designer in 1979, acting as a consultant for Alfa Romeo (1982–86) and working in collaboration with other notable Italian companies, such as Italtel and Luceplan.

Meda's work—which ranges from a luggage carousel to furniture to lighting—tends to be characterized by a lightweight minimal use of materials and an interest in transparency. His Light Light chair (1987), for example, made of carbon fiber and aluminum, weighs only two pounds.

This page and opposite

A high standard of functional performance does not have to be equated with a basic utilitarian appearance. The Berenice is a task light with a great degree of visual appeal. The refinement of the design expresses Meda's interest in minimizing material use.

Berenice (1986)

Manufacturer: Luceplan

Co-designed with Paolo Rizzatto, one of the owners of Luceplan (*see pages 58–59*), Berenice is a contemporary reworking of the classic library lamp, an antecedent that the color of its shade suggests. Meda's desire to use technology in a human, almost poetic way is revealed in the delicacy of the lamp, angled on a long, jointed arm made of dye-cast aluminum. The halogen source provides a bright, clear light with good color rendering, which adds to the functional performance.

JASPER
MORRISON b. 1959

Like his contemporary Tom Dixon, Jasper Morrison is one of a new wave of influential young British designers who use the modernist tradition as a way of responding to the possibilities offered by different production processes. Trained as a furniture designer at Kingston Polytechnic (1979–82), Morrison went on to study at the Royal College of Art in London and the Hochschule fur Kunst in Berlin. Early designs displayed the witty recycling of ready-made components; Morrison's degree show at the Royal College in 1985 was a huge critical success.

Morrison's work is characterized by a commitment to affordability and a thoughtful exploitation of manufacturing techniques. Simple, inexpensive materials such as plywood and aluminum are often used in his furniture designs, which have been produced for companies such as SCP in Britain, the German firm Vitra, and the Italian company Cappellini. The elegant Ply chair, designed for Vitra in 1988, is a case in point. Morrison has also participated in a number of international exhibitions. Other work includes a door handle for FSB, a metal and plastic bottle rack for Magis (1994), and a tram for the city of Hanover.

Glo-ball (1998)
Manufacturer: Flos

A modern classic in the making, Morrison's Glo-ball delivers a flood of warm ambient light that is softly diffused through its over-scaled etched-glass globe. Available in four different versions: a pendant lamp, two table-or floor-standing lights, and a floor lamp, Glo-ball displays a certain quality of manufacture that is clearly apparent in the weighty and durable metal base.

Like many of Morrison's furniture designs, the form of the light is spare and understated, but it is far from austere: modernism tempered with a more human sensibility. One of Morrison's aims is to produce designs that are as anonymous as ordinary household objects. Yet, while the Glo-ball is simplicity itself, its almost childlike, futuristic form has an engaging character of its own.

This page and opposite

The futuristic Glo-ball has a friendly quality—light as a companion piece. The large glass globe delivers a flood of soft ambient light.

PAUL
NEWMAN b. 1961

Born in London, Paul Newman graduated from Chelsea School of Art and Middlesex University in 1983 with a degree in furniture design and the ambition to break into the furniture manufacturing business. He first set up a small company, Bonhomi Design, with a college friend, Andrew Hilton, and together they designed and made a number of products, selling to London retailers such as Heals and Liberty. One design, for a stackable table, won a Design Council award in 1986.

At the International Furniture Fair in Milan in 1988, Newman met an Australian architect, Rob Whyte, who shared his aspirations. The result was a new partnership and the launch of Aero/UK in 1989, a company designing and manufacturing modern furniture and accessories, which were sold through Aero retail outlets in Britain and Australia. However, the economic climate was not favorable, and the recession of the early 1990s lead to a restructuring of the business. Whyte returned to Australia, and Newman established Aero Wholesale in 1991 to sell Aero products to other retailers, a move which improved their commerciality and brought eventual success. The company now includes a contract service for architects and trade specifiers, stores, and showrooms, and a mail-order business, as well as the wholesale side.

Above and opposite

These good-looking, no-nonsense aluminum hanging shades, derived from industrial light fixtures, are a perfect expression of the loft aesthetic.

Spun (1991)

Manufacturer: Aero Wholesale Ltd

The classic modernity of these simple aluminum hanging fixtures reflect the prevailing Aero philosophy: to design, manufacture, and sell good-looking, practical, and affordable contemporary furniture and fixtures for the home. Lightweight, durable, and ideal for use as a reflector, aluminum was a favorite material of early modernist lighting designers. These utilitarian pendant shades, with their industrial good looks, come in a variety of diameters and shapes, and are made using original industrial spinning tools from the 1950s.

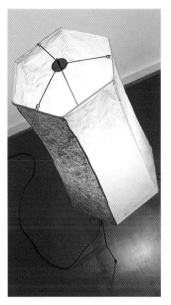

ISAMU
NOGUCHI 1904–1988

American designer and sculptor Isamu Noguchi was born in Los Angeles to an American writer and Japanese poet. His upbringing was similarly bicultural, with his early childhood spent in Japan and his school days in the United States. Noguchi studied art in New York and sculpture in Paris with the celebrated artist Constantin Brancusi at the end of the 1920s. In the early 1930s he returned to New York and began to work as a freelance designer and sculptor; early designs included stage sets for the dancer Martha Graham, as well as furniture and other products. Noguchi's bakelite radio, Nurse, was designed in 1938.

One of Noguchi's most famous designs—and one that is still in production—is a glass-topped coffee table that rests on a separate, organically shaped wooden base (1944, for Herman Miller). But he is probably best known for his paper lanterns, which he began to create in the early 1950s after spending some time in Japan. Combining both a Brancusi-like sculptural quality with the elemental form of traditional Japanese design, these lights have been widely imitated in recent years, but rarely to the same high standard of quality.

Akari paper lanterns (from 1951)

Manufacturer: Knoll

This page and opposite

The widely imitated and hugely influential Akari series derives from traditional Japanese paper and bamboo lanterns. These original Noguchi designs come in a variety of forms, from columns and cylinders to beehive shapes.

Noguchi's experimental light sculptures, which he dubbed "Lunars," led to his involvement with lighting design. These sculptures were the starting point for the Akari series of paper lanterns, designed by Noguchi for Knoll, which brought him international recognition. The first was a paper cylinder standing on three wire legs and concealing a lightbulb; later versions ranged from oversized spheres to twisted, hanging forms. The paper is handmade in Japan from mulberry bark; like traditional Japanese lanterns (or "chochins") with their bamboo frameworks, the shades are collapsible. "Akari" means light in Japanese, in both senses of the word, and Noguchi prized the paper for its "human warmth." His designs were originally produced in expensive limited editions and have been much imitated ever since.

Verner Panton was a design
iconoclast who rejected the
contemporary aesthetic of his native
Denmark in favor of synthetic
materials and space-age forms.
The Panthella may be reasonably
modest and understated for a Panton
design, but its evocative use of
acrylic and luscious curves are
entirely characteristic.

VERNER
PANTON 1926–1998

Panton's total lack of interest in the craft aesthetic or in natural materials was most unusual for a designer born and educated in Denmark. After graduating from the Royal Danish Academy of Fine Arts in Copenhagen in 1951, Panton went to work in the furniture studio of Arne Jacobsen before making a complete break with both his native country and its design traditions. After a period touring Europe, he moved to Switzerland and set up his own studio in Binningen in 1955.

Panton's rebellion took the form of a dedication to new synthetic materials and an abiding interest in ultra-modern futuristic and space-age forms. A series of chairs designed at the end of the 1950s and beginning of the 1960s exploited the latest developments in materials technology. The Panton chair (1960), perhaps his most famous and celebrated design, is a single fluid sweep of injection-molded plastic, its glossy curves echoing automotive design. It was the first one-piece cantilever chair. Panton was also a noted color theorist, famous for creative "happenings" and other manifestations of pop culture.

Panthella (1970)

Manufacturer: Louis Poulsen

Panton's preoccupation with space-age forms can be seen in the design of this table lamp with its organic curves. The mushroomlike hemisphere of the shade, made of acrylic, is supported by an almost delicate stem that flares out at the base. Both the use of modern materials and the futuristic shape serve to give the design unity. Panton was also responsible for a number of other highly original and expressive lighting designs, including oversized chandeliers made of clusters of plastic disks suspended in patterns that recall molecular structures.

PAOLO
RIZZATTO b. 1941

Trained as an architect in his native Milan, Rizzatto worked for Arteluce for a decade after graduation, before opening his own studio in 1968. His work ranges from buildings, including the town hall at Torricella Peligna, to furniture and lighting designs, produced for companies such as Arteluce, Cassina, Alias, and Molteni. In 1978 he was one of the co-founders of the lighting firm Luceplan.

Rizzatto shares an interest in the potential of technology with his sometime collaborator Alberto Meda, with whom he worked on the design of the Berenice (*see pages 48–49*) and Titania lamps. His willingness to combine different materials and styles give his designs an almost timeless quality that sets them apart. The Dakota chair, which Rizzatto designed for Cassina, for example, features an aluminum seat covered in a combination of leather and polypropylene.

Costanza (1986)

Manufacturer: Luceplan

An elegant and economical summary of the classic floor lamp, Costanza is available in a variety of different formats, including a telescopic version that allows the height of the main stem to be adjusted. While the form of the light is archetypal, the materials betray the entirely contemporary origin of the design: the wraparound shade (available in a number of different colors) is cut from polycarbonate, while the stand is sleek aluminum.

A slender stem, which juts out from the main stem at a 45-degree angle, extends below the base of the shade and acts as the light switch. In one version of the design, this switch is touch-sensive, so the light can be dimmed or brightened by degrees.

This page and opposite

The standard lamp has had an
old-fogey image until recent years,
when the basic form has been
reworked by a number of different
contemporary designers. The
Costanza is one of the more elegant
of these reinterpretations. The
thoroughly modern materials
employed in the design are
polycarbonate and aluminum.

RICHARD
SAPPER b. 1932

Born in Munich, Richard Sapper studied a diverse range of subjects from mechanical engineering to economics before joining the styling office of Daimler Benz in Stuttgart in 1956. In 1958, he moved to Milan, where he worked first in Gio Ponti's architectural office and then in collaboration with the Italian designer Marco Zanuso. Throughout the 1960s and 1970s, Sapper and Zanuso were responsible for a range of iconic designs that married German precision and rationality with Italian glamour and flair.

In 1972 Sapper returned to Munich, where he set up his own studio. The hugely successful Tizio light, designed for Italian lighting company Artemide, won him international recognition. Subsequent designs throughout the 1980s, notably the Bollitore kettle (1983) for Alessi, gained an equivalent mass exposure and established his name as a designer of cult objects. Sapper's work for Alessi, which has included a line of household products such as cooking pots and flatware, shows meticulous attention to detail. In 1981 he became product design consultant to IBM and has been responsible for a number of elegantly minimal computers, notably the Leapfrog (1992).

This page and opposite

The graphic silhouette of the Tizio lamp is a familiar feature of many contemporary interiors. Elegantly attenuated in black, it is equally striking in pure white. It is adaptable enough to be used in many different areas of the home, or wherever bright task light is required.

Tizio (1972)

Manufacturer: Artemide

The phenomenally successful Tizio, manufactured and sold in huge numbers, typified the matt-black aesthetic of the 1970s. A radical updating of the Anglepoise idea, this elegant articulated light is exceptionally responsive, thanks to a precise balancing mechanism that allows repositioning at the touch of a finger.

The minimal yet dramatic design owes at least some of its impact to the absence of any trailing wires or cables. Instead, low-voltage power is routed directly through the metal construction. Designed essentially as a desk lamp, the Tizio can also be used as a standard lamp with the arm swung in an upward position.

ROLAND
SIMMONS b. 1946

The American designer Roland Simmons was born in Cowley, Wyoming, where
he now lives and works. He was educated at Brigham Young University in Provo,
Utah, doing a first degree in industrial design before a masters in design and the
philosophy of aesthetics. It was at graduate school that Simmons first became
interested in designing in paper, attracted to the material because of its
"cleanliness" and the unique way in which light is transmitted through it.

Simmons's Lumalight, the design that arose out of these preoccupations, has
been widely exhibited. It has been selected for shows both in North America and
Europe, including "Design in Spring" Barcelona (1999), as well as a traveling
exhibition that toured the Far East at the end of 1999.

Lumalight (1972)

Manufacturer: Interfold

The statuesque Lumalight is based on concepts that Simmons first developed in graduate school. The light is now available in a choice of different formats, including wall sconces, table, and floor lamps, although the floor lamp is probably the best-known version.

A lofty indented column constructed from recyled cardboard, the Lumalight serves as a dramatic spatial marker, softly glowing with ambient light when lit, boldly sculptural when unlit. It is manufactured in Wyoming by Simmons's own company, Interfold.

This page and opposite

Simmons's Lumalight takes as its starting point the excellent diffusing properties of paper. The gentle light emitted from this lofty floor-standing column provides atmospheric background illumination; when unlit, the sculptural elements of the design come into their own.

PHILIPPE
STARCK b. 1949

Probably the best-known designer in the world today, Philippe Starck has a reputation as something of an *enfant terrible*. His work is wide-ranging—from hotel interiors to noodles, from chairs to toothbrushes—and his output is astoundingly prolific. Many of his designs, whether for furniture, buildings, or everyday objects, display an interest in zoomorphic form and an emotional, associative quality that is uniquely his own. The Starck lemon squeezer, Juicy Salif (1990, Alessi), in particular, has become a modern design icon.

Starck was born in Paris and studied furniture and interior design at the Camondo School before becoming artistic director at Pierre Cardin in 1969 at the age of only twenty. In the late 1970s he founded Starck Products, and in 1983 he was one of a group of young designers chosen by President Mitterand to remodel interiors at the Elysée Palace. It was, however, Starck's commission to design the furniture and interiors of the Café Costes in Paris the previous year that first brought him international acclaim.

Furniture designs include a series of chairs produced for companies such as Driade, Vitra, Disform, and Idee. The Royalton, New York (1988), the Peninsula, Hong Kong (1995), and the Delano, Miami (1995), are among his best-known hotel interiors. A recent preoccupation is the "Good Goods" mail-order brand, launched in 1998, a line of basic, cost-effective, and environmentally friendly products.

Rosy Angelis (1994)

Manufacturer: Flos

Standing on three spindly legs, Rosy Angelis reduces the standard lamp to its most fundamental elements. The shade is a piece of cotton fabric, simply draped over the frame to hang in soft folds. It is the contrast between the almost frail-looking tripod and the impromptu shade that gives the design its emotive power. Suggestive of an unveiling—or an old-fashioned camera— the mysterious appeal of Rosy Angelis reveals Starck's belief that design should provoke emotions in people.

This page and opposite

Starck is a designer of protean energy who has set about redefining a whole range of household objects and furnishings. His mysterious draped standard lamp, the Rosy Angelis, is a good example of his use of design as a means of arousing an emotional response.

This page and opposite

Miss Sissi, Starck's version of the café table lamp, is clever, playful, and hugely appealing. Made of technopolymer plastic, the light comes in a choice of colors.

Miss Sissi (1991)

Manufacturer: Flos

Starck's best-known lighting design and a huge commercial success, the Miss Sissi table lamp is a reworking of the classic café table light. Engaging both in its simplicity and its smallness—the light is only 11 inches high—Miss Sissi is made of colored technopolymer plastic, with a "stitched" detail on the base and stem that suggests a fabric covering. A flared ripple in the base accommodates the cord and holds it in position. The lamp provides both direct and diffused light. Many of the names Starck chooses for his designs come from the work of science-fiction writer Philip K. Dick.

Ara (1988)

Manufacturer: Flos

There was a time in Starck's career when practically all of his designs featured a "horn" motif in some form or other—as humorous flourishes on cheese dishes, as spurs on chair legs, or in the fluid curve of a door handle. This halogen table lamp is a classic example; the sleek, organic shape is emphasized by the use of polished chromium-plated metal. Instantly recognizable as a Starck design, the swept back lines of the lamp give it a certain dynamism and a quirky character. The effect is to treat lighting as an active intrusion rather than a passive background presence.

This page and opposite
Ara, a halogen table lamp made in sleek chromium-plated metal, is a design that blows its own horn.

WILHELM
WAGENFELD 1900–1990

Wilhelm Wagenfeld was a student at the Bauhaus at Weimar, studying under Laszlo Maholy-Nagy in the metal workshop, when he designed the WG 24 in collaboration with a fellow student. Little is known of his co-designer, Karl Jucker. Wagenfeld, however, went on to become a well-known industrial designer in Germany, his native country, where, unlike many of those associated with the Bauhaus, he remained throughout the World War II years.

Between 1931 and 1935 Wagenfeld taught at the college of art in Berlin and subsequently worked at a glass company until 1947; later, he became a professor at Berlin's Hochschule fur Bildende Kunste. Much of his output was in glassware (some of which is still in production today), but he also designed flatware, light fixtures, and a set of china and tray for Lufthansa in 1955. One of the most ubiquitous of his designs was a glass ink bottle for Pelikan (1938); other well-known Wagenfeld products include an ovenproof casserole dish (1938) and the Max und Moritz salt and pepper shakers (1952). From 1954 Wagenfeld had his own design studio in Stuttgart, working for companies such as Rosenthal and Braun. He remained loyal to Bauhaus teachings throughout his career, bringing modernist ideals to the design of everyday mass-produced products.

This page and opposite

Still able to hold its own more than three-quarters of a century after it was first designed, the WG 24 is evidence of the prophetic work of the Bauhaus school.

WG 24 (1923–24)

Manufacturer: Tecnolumen

The WG 24, with its industrial appearance, is a classic product of the Bauhaus school. The opalescent glass shade makes reference to a common component of factory lighting, while inside the clear glass stem, steel tubing is used to carry the wiring. The metal parts of the lamp are glossily nickel-plated. The light was displayed at the Leipzig Trade Fair in 1924, but original examples, which were hand-produced, proved prohibitively expensive for the ordinary domestic market. The WG 24 is one of the few lighting designs to come out of the Bauhaus; its director, the eminent architect Walter Gropius, chose this light for his own living quarters at the school.

Right A classic pop product, the original Astro was first designed in 1963, and is now enjoying a second burst of popularity in the wake of current interest in 1960s and 1970s culture. **Above** A variety of new colors and designs were introduced in the early 1990s.

GEORGE CRAVEN
WALKER b. 1918

The original "lava" lamp, the Astro, was invented by British-born George Craven Walker in 1963. Walker, an ex-RAF man, avid helicopter pilot, and practicing naturist, found himself fascinated by a wartime invention for an egg timer, which made use of the natural repulsion of oil and water. After seven years of trial and error, Walker came up with the notion of using a lightbulb to heat wax suspended in a bottle of water. The "lava" lamp was born.

Irreverent, fun, and kitsch, the Astro proved the perfect pop product. But by the end of the 1980s, sales had slumped. In 1990, Walker sold controlling shares in the company he had set up to manufacture the lamps to Cressida Granger, a specialist in 1960s furniture and lighting; at that time, the firm employed just four people and made 10,000 lamps a year. New colors, materials, and models were introduced, and with the increasing interest in retro design, the "lava" lamp attracted a whole new generation of fans. Sales have rocketed; Mathmos, as the new company is called, now employs 150 and produces 800,000 lamps a year.

The company now makes a range of designs as well as the original Astro. Not surprisingly, given the affection these designs inspire, there is a Mathmos collector's club for those keen to track down original and contemporary models.

Astro (1963)

Manufacturer: Mathmos

The basic idea of the lamp is essentially simple: a lightbulb in the base of the fixture heats a bottle filled with colored wax suspended in water. As the wax warms, it forms shifting globular shapes in a mesmerizing display of color and movement.

The lamp requires a certain degree of care in handling. The wax will only rise once the bottle is sufficiently heated, but overheating can lead to excessive bubbling. The lamp should be kept out of drafts, and best results occur when the room is not too hot or too cold. Moving the lamp around while it is switched on, or shaking the bottle, can lead to permanent cloudiness.

This page and opposite

The unassuming elegance of this aluminum task light is matched by its versatility. The design is also available in a wall-mounted version.

HANNES WETTSTEIN b. 1958

Hannes Wettstein was born in Ascona, Switzerland. A prolific designer whose work ranges from interior design to carpets, furniture to watches, he is the recipient of many international awards. From 1982 to 1988, Wettstein worked in collaboration with the Buro fur Gestaltung atelier in Zurich, before becoming a partner of the Design Eclat agency in Erlenbach (1989–91). His particular interest in lighting design led him to take up the post of the art director of the Italian Lighting company O Luce (1992–94); since 1993 he has been a partner of the design group 9D in Zurich, which he founded.

Wettstein's broad interests are reflected in the scope of his output: an eyewear collection, handtufted carpets, a computer support, watches, handles, audio devices, measuring instruments, a lady's razor, a chronometer, a milk glass—as well as chairs, tables, lights, and interiors. His interior design credits include the Pink Flamingo shops in Zurich and Bern (1980s), a Zurich bookstore, Folio Books and Looks (1990), and the restoration of the Swiss embassy in Madrid (1994) and the Swiss embassy in Teheran (1995). In 1995 he was also responsible for the interior design of the Grand Hyatt Hotel in Berlin.

Lighting is a central preoccupation. In 1982 Wettstein designed Metro, the first system of low tension lighting using two contact cables for Belux, Wohlen, for whom he also designed the Lucifer wall light in 1991 and the Cyos spotlight collection in 1995. Lights for O Luce include the Intro and Soiree lamps.

Spy (1996–98)

Manufacturer: Artemide

Available as a floor lamp, table lamp, or wall-mounted version, the adjustable Spy is an elegant directional light in sleek aluminum. In the floor and table lamp format, the arm can be moved on the vertical axis, while the head can rotate horizontally; the head of the wall-mounted design can also be repositioned. The same flexibility extends to compatible light sources: Spy can be used in conjuction with incandescent, halogen, fluorescent, or energy-saving bulbs.

PETER
WYLLY b. 1967

British designer Peter Wylly graduated from Middlesex University in 1989 with a degree in fashion design. But his work had attracted some attention even before graduation, when he was chosen as a finalist in the Crespi Plastic design awards in Milan and his student home was featured in *World of Interiors*.

After leaving college, Wylly started a business making exhibition stands and tent structures, which led on to his first wrapped lamp prototype. After spending six months in Siena, Italy, building a tree house, he returned to Britain to set up Peter Wylly Lighting and put the wrapped lamps into production.

Wylly's wrapped lamp collection was an immediate success. The Conran Shop and Heals bought the collection, while Romeo Gigli specified the designs in five new stores worldwide. This was followed, in 1996, by a second lighting collection comprising the Barbara ceramic lamp, Eclipse plastic lamps, woven lampshades, and sculptural wrapped pieces. Babylon Design Limited was launched in 1997 by Wylly and his partner, Birgit Israel; since then, Wylly's work has ranged from lights in a variety of different materials—including ceramic, paper, and polypropylene—to ceramic accessories. In 1999, his China table lamp won the Design Plus award.

DNA (1997)
Manufacturer: Babylon Design Limited

Left and above Like the floor lamp, the central pendant fixture or chandelier has undergone considerable reinvention in recent years. Wylly's DNA design, with its molecular theme, updates the chandelier for contemporary living.

The traditional form of the chandelier has proved a point of departure for contemporary lighting designers. With its multiple points of light and sculptural potential, the chandelier represents the most appealing form of central fixture and adds drama and a sense of unity to a room.

Wylly's DNA design is a flexible structure in clear polypropylene that can be assembled in various ways. The standard chandeliers are either six- or nine-piece, but an 18-piece contract chandelier can also be produced to order. Small crown-silvered bulbs avoid the problem of glare—always an issue in overhead lighting—while the looping bands of polypropylene serve both to diffuse the light and suggest the double helix spiral of DNA.

RETAILERS AND MANUFACTURERS

2 Danes
73 White Bridge Road
Suite 109
Nashville, TN 37205
615-352-6085
email:2danes@netware.net
European contemporary lighting.

Architectural Lighting Resources, Inc.
11042 Elm Street
Omaha, NE 68144
402-397-2867
European and architectural lighting products.

Arkitektura IN-SITU
474 North Old Woodward
Birmingham, MI 48009
248-646-0097
Modern classic lighting and furniture.

Baldinger Architectural Lighting
19-02 Steinway Street
Astoria, NY 11105
718-204-5700
www.baldinger.com
*Custom and decorative lighting. Sole distributor
of Best & Lloyd lamps in North America.*

Cantoni
4800 Alpha Road
Dallas, TX 75244
972-934-9191
888-226-8664
www.cantoni.com
Modern furniture, lighting and accessories.

City Schemes
1050 Massachusetts Avenue
Cambridge, MA 02138
617-497-0707
www.cityschemes.com
Contemporary European furniture.

DIVA Inc.
8801 Beverly Boulevard
Los Angeles, CA 90048
310-278-3191
Contemporary Italian furniture.

Domus
1919 Piedmont Road
Atlanta, GA 30324
404-872-1050
800-432-2713
email:sales@domusinternational.com
www.domusinternational.com
*High-end European contemporary furniture,
lighting and accessories.*

Form + Function
1300 East 86th Street
Indianapolis, IN 46240
888-722-9992
Contemporary and modern classic lighting.

Hinson & Co.
Hansen Lamps
979 Third Avenue
New York, NY 10022
212-688-7754
*They specialize in the original swing-arm wall
lamp by George Hansen.*

Illuminating Experiences
233 Cleveland Avenue
Highland Park, NJ 08904
732-745-5858
Contemporary and classic lighting.

Lee's Studio
1755 Broadway
New York, NY 10019
877-LIGHT-NY
Contemporary Italian lighting.

Lighting Unlimited
4025 Richmond Avenue
Houston, TX 77027
713-626-4025
800-447-5240
email:lightpros@aol.com
Contemporary lighting and light planning.

Limn
290 Townsend Street
San Francisco, CA 94107
415-543-5466
www.limn.com
European lighting.

Lumature
15620 N. Scottsdale Road
Scottsdale, AZ 85254
480-998-5505
www.lumature.com
Contemporary European lighting.

Luminaire
2331 Ponce De Leon Boulevard
Coral Gables, Fl 33134
305-448-7367
and at
301 West Superior
Chicago, IL 60610
312-664-9582
www.luminaire.com
Brilliant European design at home and work.

Lunatika
900 Lincoln Road
Miami Beach, Fl 33139
305-534-8585
www.lunatika.com
European contemporary lighting.

Lux Lighting Inc.
1109 NW Glisan
Portland, OR 97209
503-299-6754
Contemporary lighting.

Mecca Modern Interior
21A South Broadway
Denver, CO 80209
888-307-2600
www.meccainterior.com
*Modern and Post-modern American and European
design furniture and accessories.*

The Morson Collection
100 East Walton Street
Chicago, Il 60611
312-587-7400
and at
31 St James Avenue
Boston, MA 02116
617-482-2335
*High-end contemporary furniture and lighting .
Call 800-204-2514 if located outside of Illinois
and Massachusetts.*

OLC
152 North Third Street
Philadelphia, PA 19106
215-923-6085
Contemporary lighting and furniture.

Sheehan's Office Interiors
524 Park Avenue
P.O. Box 629
Portsmouth, RI 02871
401-683-3150
www.sheehansoffice.com
Interior design and contract furniture.

Source Of Light
229 N. Damen Avenue
Chicago, IL 60612
312-421-5841
Contemporary state-of-the-art lighting.

Tech Lighting Galleries
300 W. Superior
Chicago, IL 60610
312-944-1000
www.tlgalleries.com
Specialists in monorail, twin rail and cable lights.

Totem Design Group
71 Franklin Street
New York, NY 10013
212-925-5506
888-519-5587
www.totemdesign.com

Voltage
2703 Observatory Avenue
Cincinnati, OH 45208
513-871-5483
email:voltage@one.net
www.voltageinc.com
European contemporary lighting and accessories.

MANUFACTURERS:

Aero
www.aero-furniture.com

Anglepoise
www.anglepoise.com

Artemide Inc.
1980 New Highway
Farmingdale, NY 11735
516-694-9292
www.artemide.com

Babylon Design Limited
+44 20 7729 3321

Best & Lloyd
+44 121 558 1191

ClassiCon
www.classicon.com

Eurolounge
www.eurolounge.co.uk (this web site
should be up and running in a few weeks)

Flos Inc
200 McKay Road
Huntington Station, NY 11746
516-549-2745

Ingo Maurer
89 Grand Street
New York, NY 10013
212-965-8817
www.ingo-maurer.com

Interfold
1 East First South
P.O. Box 121
Cowley, WY 82420
307-548-7670
www.rolandsimmons.com

Knoll
105 Wooster Street
New York, NY 10012
212-343-4000
www.knoll.com

Le Klint
www.leklint.com

Luceplan USA
315 Hudson Street
New York, NY 10013
212-989-6265

Lumina
www.lumina.it

Mathmos
www.mathmos.com

Metalarte
+34 93 477 00 86

Pallucco Italia
www.palluccobellato.it

Poulsen Lighting Inc.
3260 Meridian Parkway
Ft. Lauderdale, FL 33331
954-349-2525
email:info@louispoulsen.com
www.louis-poulsen.com

ARCHITECTS AND DESIGNERS whose work has been featured in this book:

AEM
80 O'Donnell Court
Brunswick Centre
Brunswick Square
London WC1N 1NX
+44 20 7713 9191
fax: +44 20 7713 9199
*Pages 6-7, 11 r, 27, 30, 31, 44-45,
56-57*

Avanti Architect Limited
1 Torriano Mews
London NW5 2RZ
+44 20 7284 1616
fax: +44 20 7284 1555
Pages 14-15, 32-33, 37 t

Babylon Design Ltd
Unit 7, New Inn Square
8/13 New Inn Street
London EC2A 3PY
+44 20 7729 3321
Pages 11 l, 62, 63

Claire Bataille & Paul ibens
Vekestraat 13 Bus 14
2000 Antwerpen
Belgium
+32 3 213 86 20
Page 48 b

Bilhuber Inc.
330 East 59th Street
6th Floor
New York, NY 10022
USA
212 308 4888
Page 54 b

Laura Bohn Design Associates
30 West 26th Street
New York, NY 10011
USA
212 243 1935
Pages 26 t & b, 42 b

Brookes Stacey Randall
New Hibernia House
Winchester Walk
London SE1 9AG
+44 20 7403 0707
fax: +44 20 7403 0880
email:info@bsr-architects.com
Pages 16-17

Hudson Featherstone Architects
49-59 Old Street
London EC1V 9HX
+44 20 7490 5656
Page 8

Littman Goddard Hogarth
12 Chelsea Wharf
15 Lots Road
London SW10 0QJ
+44 20 7351 7871
fax: +44 20 7351 4110
Pages 28, 29, 31, 36, 37 b, 46

Jonathan Reed
Reed Creative Services Ltd
151a Sydney Street
London SW3 6NT
+44 20 7565 0066
fax: +44 20 7565 0067
Page 3 r

Jim Ruscitto
Ruscitto, Latham, Blanton
PO Box 419
Sun Valley
Idaho, ID 83353
USA
Pages 60-61

Charles Rutherfoord
51 The Chase
London SW4 0NP
+44 20 7627 0182
Page 20

Guy Stansfeld
+44 20 7727 0133
Page 12 b

Stickland Coombe Architecture
258 Lavender Hill
London SW11 1LJ
+44 20 7924 1699
Page 28 inset

Urban Salon Ltd
Unit D
Flat Iron yard
Ayres Street
London SE1 1ES
+44 20 7357 8800
fax: +44 20 7407 2800
*Pages 10, 18, 18-19, 24-25, 52-53,
70-71*

Will White
326 Portobello Road
London W10 5RW
+44 20 8964 8052
Pages 20-21

Vicente Wolf Associates, Inc.
333 West 39th Street
New York, NY 10018
USA
212 465 0590
Page 19

Woolf Architects
39-51 Highgate Road
London NW5 1RT
+44 20 7428 9500
Page 47

ACKNOWLEDGMENTS

All photographs are by Chris Everard unless otherwise stated

Key: **t** = top, **b** = below, **l** = left, **r** = right, **c** = center

Endpapers photographer Ray Main, Babylon Design; **1** Reuben Barrett's apartment in London, light courtesy of Mathmos; **2** Jonathan Wilson's apartment in London, light courtesy of SCP; **3 l** light courtesty of London Lighting; **3 c** light courtesy of Geoffrey Drayton; **3 r** photographer Ray Main, Jonathan Reed's apartment in London; **4** photographer James Merrell; **6** photographer Ray Main; **6–7** a loft apartment in London designed by AEM Architects, from left to right lights courtesy of Viaduct, London Lighting and Skandium; **8** photographer Henry Bourne, Baggy House designed by Hudson Featherstone Architects; **8–9 & 9** lights courtesy of London Lighting; **10** Simon Crookall's apartment in London designed by Urban Salon, light courtesy of SCP; **11 l** photographer Ray Main, Babylon Design, light courtesy of Artemide; **11 tc** Christina Wilson's house in London; **11 r** a loft apartment in London designed by AEM Architects; **12 t** photographer Andrew Wood, Century Design; **12 b** photographer Andrew Wood, a house designed by Guy Stansfeld; **12–13** Christina Wilson's house in London, light courtesy of SCP; **13r** photographer James Merrell, Ian Bartlett & Christine Walsh's house in London; **14–15** Justin de Syllas & Annette Main's house in London, light courtesy of Artemide; **16 l** photographer Andrew Wood; **16–17** photographer Andrew Wood, Freddie Daniells' loft apartment in London designed by Brookes Stacey Randall; **18 & 18–19** Simon Crookall's apartment in London designed by Urban Salon, light courtesy of Geoffrey Drayton; **19** photographer Ray Main, Vicente Wolf's house on Long Island, lighting from Lightforms; **20** photographer Henry Bourne, Charles Rutherfoord's house in London; **20–21** photographer Tom Leighton, Siobhan Squire & Gavin Lyndsey's loft in London designed by Will White; **22** Christina Wilson's house in London; **23** Glenn Carwithen & Sue Miller's house in London, light courtesy of London Lighting; **24–25** Simon Crookall's apartment in London designed by Urban Salon, light courtesy of Geoffrey Drayton; **26 t & b** photographer Ray Main, an apartment in New York designed by Laura Bohn Design Associates Inc., lighting from Lightforms; **27** a loft apartment in London designed by AEM Architects, light courtesy of Mr Light; **28 & 29** an apartment in London designed by Littman Goddard Hogarth, light courtesy of SCP; **28 inset** photographer Andrew Wood, Anthony Swanson's apartment in London designed by Stickland Coombe Architecture; **30 & 31** a loft apartment in London designed by AEM architects light courtesy of Viaduct; **32–33** Justin de Syllas & Annette Main's house in London; **34–35** photographer Ray Main; **36 & 37 b** an apartment in London designed by Littman Goddard Hogarth, light courtesy of Skandium; **37 t** Justin de Syllas, light courtesy of Skandium; **38–39** Glenn Carwithen & Sue Miller's house in London, light courtesy of Coexistence; **40 & 41** Christina Wilson's house in London, lights courtesy of Century; **42 b** photographer Ray Main, an apartment in New York designed by Laura Bohn Design Associates Inc.; **43 r** photographer James Merrell; **44–45** a loft apartment in London designed by AEM architects, light courtesy of Artemide; **46** an apartment in London designed by Littman Goddard Hogarth, light courtesy of Atrium; **47** photographer Andrew Wood, Patricia Ijaz's apartment in London designed by Woolf Architects; **48** Christina Wilson's house in London, light courtesy of Coexistence; **48 b** photographer Andrew Wood, an apartment in Belgium designed by Claire Bataille & Paul ibens; **50–51** Jonathan Wilson's apartment in London, light courtesy of SCP; **52–53** Simon Crookall's apartment in London designed by Urban Salon; **54 t** Reuben Barrett's apartment in London, light courtesy of Coexistence; **54 b** photographer Ray Main, a house in Pennsylvania designed by Jeffrey Bilhuber; **54–55** light courtesy of Coexistence; **55** photographer Ray Main, photographed courtesy of Tsé & Tsé associées, Catherine Levy & Sigolène Prébois; **56–57** a loft apartment in London designed by AEM Architects, lights courtesy of Skandium; **58 l** Jonathan Wilson's apartment in London; **60–61** photographer James Merrell, a house in Idaho designed by Jim Ruscitto; **61** photographer Ray Main; **62** photographer Ray Main light from Babylon Design; **63** photographer Andrew Wood, Babylon Design; **64–65** Reuben Barrett's apartment in London, light courtesy of Purves & Purves; **66–67** photographer Ray Main; **66 inset** light courtesy of Geoffrey Drayton; **67** light courtesy of London Lighting; **68–69** light courtesy of London Lighting; **70–71** Simon Crookall's apartment in London designed by Urban Salon, light courtesy of London Lighting; **72** photographer Ray Main; **72–73** Reuben Barrett's apartment in London, light courtesy of Mathmos; **74 & 75** Reuben Barrett's apartment in London, light courtesy of Coexistence; **76 & 77** Andrew Wood, lights from Babylon Design; **80** light courtesy of Coexistence.

The author and publisher would like to thank everyone who allowed us to photograph their homes and all the retailers and manufacturers who loaned us items for photography. Special thanks to Annabel at Coexistence and Christina at Skandium. The author would like to thank Annabel Morgan, Ashley Western, Kate Brunt and Chris Everard for their creative efforts, enthusiasm and skill; with special thanks to Annabel for having the idea in the first place.